Six-Word Lessons for
A TRUSTWORTHY DRONE BUSINESS

100 Lessons to Optimize the Drone Entrepreneur/ Client Relationship

2nd Edition

Dr. Tulinda Larsen
Ruth Blomquist

Published by Pacelli Publishing
Bellevue, Washington

Six-Word Lessons for a Trustworthy Drone Business

All rights reserved. No part of this book may be reproduced or transmitted in any form or by any means, electronic or mechanical including photocopying, recording or by any information storage or retrieval system, without the written permission of the publisher, except where permitted by law.

Limit of Liability: While the author and the publisher have used their best efforts in preparing this book, they make no representation or warranties with respect to accuracy or completeness of the content of this book. The advice and strategies contained herein may not be suitable for your situation. Consult with a professional when appropriate.

Copyright © 2018, 2021 (2nd Edition) by Tulinda Larsen and Ruth Blomquist

Published by Pacelli Publishing
9905 Lake Washington Blvd. NE, #D-103
Bellevue, Washington 98004
PacelliPublishing.com

Cover and interior designed by Pacelli Publishing
Author photo by Susan Casley

ISBN-10: 1-933750-89-8
ISBN-13: 978-1-933750-89-7

Dedication

This book is dedicated to my husband, Carl Larsen, my drone catcher and business strategist.

Testimonials

"Dr. Larsen's book should be your guide for considerations as you look to start a drone service business or to hire a commercial drone service." - Sharon Rossmark, Founder/CEO, Women And Drones

"Dr. Larsen's expertise has given us new tools to help monitor and protect our environment and to document conditions on an important historic property. Her book is an essential tool for anyone hiring a drone company." – Eric Swenson, Executive Director, Hempstead Harbor Protection Committee

"We are finding drone businesses are entering the market offering many services. Dr. Larsen's book provides guidance for buyers and sellers of drone services." - Bob Tagart, Publisher, *Old Town Crier*, Alexandria, Virginia

"I recently started a multi-purpose drone business to augment my long-established drone usage in real estate. Dr. Larsen's book has proven to be a very fine resource as my business expands into other drone disciplines." – Kevin Brennan, Owner, Alfa Drones 4U

Contents

Introduction 6
Acknowledgements 8
Introduction to the Growing Drone Business 9
Examples of Commercial Projects Using Drones 15
Choosing the Right Drones for Projects 37
Choosing the Right Sensors for Projects 45
Understanding Types of Drone Service Companies 57
Understanding Important FAA Commercial Drone Regulations 65
Best Practices for Safe Drone Operations 87
Importance of Advanced Drone Operations Training 95
Preparing Bids for Government Drone Contracts 101
Finding Drone Companies and Selling Services 109
Contact Tulinda 126
References 127
About the Six-Word Lessons Series 128

Introduction

My second edition provides updates based on new FAA rules for drone operations.

Drones, also known as unmanned aerial vehicles, are rapidly becoming indispensable business tools for many industries. The technologies surrounding drones and sensors are innovating at warp speeds. Safety features, reliability, control distance, and battery life are also rapidly improving.

Prior to the development of drones, aerial remote sensing and photography required the very expensive rental of fixed-wing or helicopter manned aircraft. With the advent of low cost drones equipped with high resolution cameras, taking aerial images is now easily affordable for all types of businesses. Images and video taken from the air offer a perspective that cannot be matched from the ground. Drones can safely operate at much lower altitudes, in more confined spaces than aircraft, and can go places unsafe for humans.

The lessons in this book come from my experiences starting my own small drone business and conducting 100 interviews with drone operators across the United States to learn what does and doesn't work.

If your company has decided to hire a drone operator, this book will help you understand the best and worst projects for drones, the best drones and sensors for the job, the government regulations, and information to help you work with drone operators.

If you are a drone operator and you have decided to create a new business, this book will help you understand what customers are looking for, the best drones and sensors for the job, navigating the government regulations, best practices for operating a drone company and information to help you work with drone customers.

Acknowledgements

For this second edition, I thank Ruth Blomquist who worked with me to update the sections. Ruth came to me as a high school intern when I was Executive Director of Utah's Deseret UAS. As part of the internship, I taught her to fly drones and to ultimately pass her Part 107 Remote Pilot in Command test. She went on to form a drone club for her high school and held a drone day for Utah Girl Scouts. At the writing of this book, she is now a sophomore at Utah State University and is an active drone pilot with the USU AggieAir program. She is also part of the team for ElectraFly, the personal flying machine.

Women And Drones (WomenandDrones.com) is an inspiration for women getting into the drone business, and I thank Sharon Rossmark and Wendy Erikson for their vision to create an organization to promote and recognize women in the commercial drone industry.

I thank Darryl Jenkins for introducing me into the amazing world of drones, for which I am most grateful.

Introduction to the Growing Drone Business

1

Drones are fast-growing small businesses

Drones came into the consumer market around 2010 with improvements in miniaturization technology, making it possible to manufacture reasonably priced drones. In 2015, the U.S. government started requiring all drones weighing more than one-half pound and less than 55 pounds to be registered. Today 2 million drones are registered, with 500,000 in commercial operations. For perspective, it took 100 years for the fleet of manned aircraft to reach 250,000.

Understanding the definition of a drone

They go by many names--Unmanned Aerial Systems (UAS), Unmanned Aerial Vehicles (UAV), quadcopters, multi-rotor aircraft, but more commonly they are known as drones, for the sounds they make when they are in flight. Drones are remotely piloted aircraft and come in a wide range of sizes from as small as the palm of your hand to high-flying military aircraft.

The magic date: August 29, 2016

On August 29, 2016, the rules governing commercial drones went into effect and created the opportunity to start new drone businesses. FAA updated commercial drone regulations January 15, 2021, the final rules were published in the Federal Register on March 10, 2021 delaying the effective date from March 16, 2021 to April 21, 2021.

4

What you'll learn from this book

If you are considering hiring a drone operator, this book will teach you things you should consider. If you are ready to launch a drone business, you will learn what you need for a successful business. The lessons in this book will provide examples of the best types of projects for drones, will help you choose the right drone for your projects, and will discuss the rules and regulations for commercial drone operations.

Examples of Commercial Projects Using Drones

Contracting or providing commercial drone services

Whether you are considering hiring a drone pilot or becoming a commercial drone pilot, it is best to start with understanding what types of projects drones can and cannot perform. The following lessons include examples of the best and worst commercial drone projects.

Cinematography for movies and TV shows

The first commercial drone operations approved by the FAA were for making movies. In 2014, the FAA approved authorization for film and TV production, which allowed Tom Cruise to star in *Top Gun 2* alongside drones. Drones are perfect for getting those amazing aerial shots and close shots on tight sets.

Realtime news coverage benefits from drones

Drones go places reporters either can't get to or would be unsafe to go. Drones film events such as accidents, disasters, conflicts and other incidents in dangerous places. Remember, don't fly a drone without permission in emergency situations. For example, some drone pilots have caused trouble for firefighters, in some cases creating conflicts with firefighting aircraft and firefighters.

Drones help make real estate sales

Homes and commercial real estate are much easier to sell when buyers can get a view from above and a 360-degree view of the property. Real estate agents trying to sell residential real estate can include amazing drone footage, along with interior and exterior shots, to intrigue and attract prospective buyers. Commercial real estate aerial shots can capture the overall property and buildings.

Hotel and resort marketing with drones

When using drones for aerial shots, resort or hotel marketing teams can show the luxurious pool, the palm trees swaying in the breeze, and the beautiful beaches. The aerial views made possible by drone photography show dynamic and unique views of resort properties that further entice customers to book their meetings and vacations.

Sports and action magazines use drones

While GoPro and other action sports video cameras have mastered the first-person point of view in sports photography, drones can capture the overhead, bigger picture view. Organizers of sporting events and competitions can hire commercial drone operators to capture great footage of everything from a skier jumping off a cliff, to marathon racers rounding the last bend before the finish line, to tracking boat races or footage of remote bike races.

Special wedding day memories from above

At weddings, a drone operator can provide extraordinary pictures and videos from above of friends and family. Just think about the breathtaking beauty of the bride and groom from above with a 360-degree video of the venue. Wedding photographers are expanding their portfolios to include the hiring of commercial drone pilots or adding drone photography to their services.

Drones for inspections safer than humans

Drones can be used effectively for inspection purposes because aerial shots provide a larger field of view and no humans are be endangered during the process. When hiring or offering drone services for specialized inspections, the drone pilot needs to understand what is being inspected for safety and best results.

Safer, cheaper, better for infrastructure surveys

Aerial surveys are used in cartography, topography, feature recognition, archaeology and GIS applications, providing information on terrestrial sites that are often difficult, or even impossible, to see or measure from the ground. Drones offer photogrammetric mapping and ortho photography that is safer for humans, far less expensive than aircraft, and produces better-quality images.

14

Power line and other cable inspections

Drones are great for inspecting power lines and other cables, such as ski lifts. Companies are using drone inspection services for tower antenna inspection, wind farms, and cell phone towers. The drone images provide assurances that everything is in working order, plus they can inspect for damage after a storm.

Aerial facility inspections can include rooftops

Drones provide aerial views of facilities, including rooftops, for inspections of damage or routine inspections. Drones can produce maps and 3D models to quickly and accurately inspect and measure roofs. Adding thermal imaging camera photos or videos will reveal potential problems with equipment overheating, which supports better preventive maintenance.

16

Solar farms better inspected by drones

The growing number of solar farms that harness the sun's energy is increasing the need for inspections. Drones can fly low and slow and inspect each of the panels in the solar farm for any damage. This information gathered from the images provides information needed to correct any problems or issues with the solar panels.

Conservation and environmental regulation compliance monitoring

Federal, state and local government programs encourage conservation and environmental regulations. Drones are used in the field by biologists and researchers to count wildlife such as birds and polar bears. Drones can find evidence of illegal logging, harmful substance dumping and more. Read *Six-Word Lessons for Drone Pilots and Outdoor Enthusiasts* for best practices in flying over nature.

Maps using drone images are powerful

Map makers, topographers and even archaeologists need aerial surveys to fully evaluate some of their projects. Drone operators can provide these aerial views with lower overhead costs. Drones can fly low and slow, which captures better images of a site than a traditional manned aircraft. Drone operators can take advantage of many mapping software options to produce mosaic or composite photos based on stitching together many images.

Drone mapping helps manage construction projects

High-resolution drone imagery is one of the most efficient and cost-effective construction tools available. From early in the bid process to project completion, drones provide needed mapping, surveying, monitoring and imaging tools, such as pre-bid job site high-resolution inspection, aerial photos and videos, progress photos and videos on active construction sites, and asset management. They can track machinery, stockpile volume and other assets from remote locations with 3D imaging and modeling.

Drones are essential for disaster relief

The hurricane season of 2017 proved that drones are essential for disaster inspections, recovery, and relief. Drones can go into areas that are unsafe for rescue workers to search for victims and assist in recovery. Drones can carry small payloads and drop in supplies, such as water. Aerial drone imagery provides the tools for insurance companies to inspect for damage and pay claims promptly.

Insurance claim drone use is growing

Insurance companies are employing drone operators to inspect and quantify claims like hail damage to roofs and weather damage to agricultural fields. Drones can provide better images and can make the work of adjustors more efficient and safe, while providing better quality data.

Drones assist and support emergency response

Equipping drones with infrared night vision sensors allows search and rescue teams to detect missing individuals by heat emissions as well as visually and drones can deliver emergency supplies as needed. Fire departments and law enforcement agencies are quickly adopting the use of drones to gather intelligence and to go into situations too dangerous for humans.

Agriculture benefits from drone aerial images

Before drones, farmers had two choices for gathering aerial images of the fields: satellites and manned aircraft. Drones are cost-effective, and have slow and low tools to collect extremely high-resolution aerial images of crops and animals. New technology software takes this data and creates maps and analyses of crop health to help farmers better manage costs and yields.

Drone racing business is fast-paced

Drone racing is a rapidly growing sport around the world. Racing drones are very small and compete on speed and accuracy through an obstacle course, either indoors or outside. The competitors put on goggles and virtually get inside the drone with a First Person View (FPV). Even ESPN covers drone racing. You can host a drone race or provide needed services to support drone racing.

Projects not good for drone operations

Drones are amazing technologies, but they can't do everything. Under current regulations, drones must be kept within the visual line of sight of the pilot or visual observer which is roughly .8 miles. Weather limits drone operations, especially wind, rain, snow, and fog. Rules on flying over people limits some projects. Drones should not be flown during emergencies, unless special authority has been granted.

Choosing the Right Drones for Projects

Drones have incredible built-in technology

Drones now have 4K cameras that can take high resolution images and video. They also have built-in GPS and autopilot, enabling them to fly preprogrammed routes and Bluetooth capability that allows live-streaming. In addition, drones have collision avoidance, terrain follow (the ability to maintain a consistent height while flying over uneven terrain), active tracking (automatically follow a selected subject), the ability to fly at speeds of 50 mph (80 km/hr) and powerful data analytic software.

Price ranges for purchasing right drone

You can buy a drone for less than $100, but they are considered toys. When investing in a drone for business, you will need to spend more than $500, but more typically between $2,000 and $5,000, plus the cost of cameras and sensors. When budgeting for your drone business, plan for repairs and replacement drones, because everyone crashes. I have lost two drones. One was replaced and the other repaired.

Multirotor copters are most popular drones

Drones can have 4, 6, 8, or even more rotors. The most popular is the 4-rotor quadcopter. Affordable multirotor drones are the result of advances in technology, including cheap lightweight flight controllers; LiPo batteries, based on Lithium Polymer chemistry, which provide very high energy density compared to other types of batteries; GPS guidance; accelerometers (IMU); and a wide range of sensors.

Fixed wings fly longer and farther

Fixed wing drones are aerodynamically able to fly longer, farther and higher than quadcopters, but regulations that require keeping the drone within a visual line of sight limit the usefulness of fixed wing drones. Take-offs of fixed wing drones require catapulting. Landings are controlled crashes, leading to damage to the drone.

Transition operation drones have unique capabilities

The third type of commercial drone has two types of operations, which allows the drone to take off vertically, then transition to horizontal flight. Transition operation drones have the benefits and safety of vertical take-off and hover, plus the better aerodynamics of fixed-wing aircraft. Plus the transition drone landings are similar to a multirotor copter and much more controlled than a fixed-wing drone.

Type of drone depends on mission

Drones can carry different payloads, cameras, sensors, and equipment. It makes sense that larger drones can carry heavier payloads. Some can even carry multiple sensors, such as two different types of cameras. However, drones with a single high-resolution camera are adequate for most drone projects.

Choosing the Right Sensors for Projects

Cameras are the most popular sensors

Cameras are the most popular visual sensors. Static photos are expressed in megapixels (12MP, 16MP) that define the number of pixels (length x height) that compose a photo. Videos are usually expressed with terms such as Full HD (1080 x 1920 pixels with 30 frame per second) or Ultra HD (3840 x 2160 pixels with 30/60 frame per second). 4K refers to horizontal screen display resolution in the order of 4,000 pixels.

Basic cameras capture red, green, blue

RGB (red, green, and blue) can be combined in various proportions to obtain any color in the visible spectrum. Levels of R, G, and B can each range from zero to 100 percent of full intensity. Basic cameras are equipped with a standard RGB sensor, through which the colored images of persons and objects are acquired.

Multispectral sensor cameras capture nonvisible light

Multispectral sensors can capture near-infrared radiation (NIR) and ultraviolet light invisible to the human eye. Multispectral sensors are instrumental in plant health and management. They can pinpoint nutrient deficiencies, identify pest damage, optimize fertilization and assess water quality. Detailed insight can also be gained using algorithms from vegetation indices.

35

Hyperspectral sensors detect the electromagnetic spectrum

Hyperspectral sensors collect hundreds of narrow bands of data along the electromagnetic spectrum. These sensors can detect and identify minerals, vegetation and other materials that are not identifiable by other sensors. They are used in plant nutrient status, plant disease identification, water quality assessment, foliar chemistry, mineral and surface chemical composition and spectral index research.

Thermal sensors measure relative heat images

Thermal sensors measure the relative surface temperature of land and objects beyond the scope of human vision. Drone-based thermal imagery has many applications including facility inspections, surveillance and security, searching for lost people, night operations, water temperature detection, water source identification, livestock detection and many other heat signature detection applications.

LiDAR sensors use lasers to survey

Light Detection and Ranging (LiDAR) sensors use light energy, emitted from a laser, to scan the ground and measure variable distances. LiDAR creates elevation data which produces high-resolution maps and 3D models of natural and man-made objects. Drone-based LiDAR collection yields very high-fidelity data. The point cloud generated from drone-based LiDAR can yield 100 to 500 points per square meter at a vertical elevation accuracy of 2 to 3 centimeters.

Sniffers can detect many gas types

Sniffer sensors can detect gasses, including methane, carbon dioxide, hydrogen fluoride, ammonia, hydrogen chloride, hydrogen cyanide, acetylene, etc. These sensors are used to find leaks and other gas emissions. The sniffer sensor analyzer will register the reading and if the concentration exceeds the programmable set point it will sound an alarm so the drone pilot can circle around and survey the leak further.

Pricing sensors and understanding processing software

The sensors described previously have a wide range of prices from hundreds to thousands of dollars. Some can be rented for special missions. The higher tech sensors have expensive post-processing software to analyze the images and data gathered. When designing a mission for a specific project, the cost of the sensor and related software needs to be considered.

Selecting image editing software for projects

Some drones come with very simple video editing right in the control software. To do really cool things with drone videos, like special effects and adding music, you need video editing software. iMovie or FinalCut are good for Macs, while Adobe Premier Pro CC and Corel VideoStudio work for PCs. Mapping software processes large numbers of images to create a single image. There are many affordable commercial mapping software packages.

Livestreaming and posting your drone videos

Many drones can livestream video to Facebook, YouTube and Vimeo. Posting edited videos to YouTube is a great way to share drone images. YouTube does have size and length limitations, but you need to have Wi-Fi to transmit the signal to livestream. I have used my phone's hot spot, but in remote locations, the signal may be too weak to transmit.

Understanding Types of Drone Service Companies

Drone companies are small and large

In the U.S., small businesses comprise 99 percent of all employer firms, employ nearly half of the workforce, and account for more than 60 percent of the private sector's net new jobs. Drones are creating opportunities for entrepreneurs to create new small businesses. Drone companies can also be very large and offer multiple services. Several drone companies have received investments approaching $100 million to provide comprehensive services to the market.

Adding to an existing business service

Drones can expand the business offerings of several types of companies. For example, established photographers can add aerial images to their portfolios, espccially wedding photographers, surveyors can add drone images and mapping to their services, crop consultants can add drones to their analysis of farms, and even marine biologists can add drones to their environmental business.

No previous aviation operations experience required

I surveyed 100 drone operators and found that 80 percent of these drone small businesses did not come from an aviation background. An example is Alimosphere (Alimosphere.com), a drone business started by a marine biologist to help drone pilots create flight plans that benefit humanity and respect wildlife. Check out *Six-Word Lessons for Drone Pilots and Outdoor Enthusiasts*.

Exploring drones as service franchise opportunities

Some companies offer franchise opportunities for drone operators. For a franchise fee, these companies provide the support services, including help with navigating the regulatory environment, standardized products, marketing for drones as service providers, and sales leads.

Part of a large drone company

Over the past decade, drones as service companies have entered the market and have grown into multi-million dollar companies. Large drone companies provide a number of services to governments and industry, including operating the drone, proprietary processing of the images from the drone, and customized recommendations from the analysis.

Companies with internal drone services

Companies in the energy, mining and construction industries have created internal drone services departments. These companies purchase drones, hire licensed pilots, manage the fleet, and have software to process the data from the drone flights. Other well-known companies with an internal drone business are Amazon, WalMart and Google, where they are developing small package delivery services using drones.

Understanding Important FAA Commercial Drone Regulations

# Commercial drone operations FAA regulatory definitions

The FAA defines commercial drone operations as, "Flying for work, business, non-recreational reasons, or commercial gain." This typically includes flying a drone for hire, compensation, to provide a service, or for economic benefit of an entity or person. Intended use, not compensation, is the determining factor.

… 49

References to FAA commercial drone rulings

Regulation of commercial drone operations can be found in the Code of Federal Regulations (CFR) Section 16 Part 107. In 2016, the FAA issued guidance on operating under Part 107 in a 600+ page ruling, which can be searched for specific operations, such as flights over people. In January 2021 these regulations were updated to cover flights over people, night operations, remote ideintification and include 300 pages.

50

Part 107 Remote Pilot in Command

Drones operated commercially are considered FAA regulated aircraft operations under Part 107 of the FAA regulations, and require a license. A Part 107 Remote Pilot In Command (RPIC) license is earned by passing the FAA Remote Pilot Knowledge Test and taking the training. In addition, the pilot must be 16 years of age and pass TSA security vetting. If a drone is being flown for fun, no licensing is required.

51

High penalties if not FAA licensed

The penalty for operating a drone commercially, without a Part 107 Remote Pilot License, is $1,100 per violation for the pilot and $11,000 for the organization. A single flight could have multiple violations. Make sure every drone pilot you hire is licensed. The test is not difficult and there are several online prep courses. If you are thinking of charging for your services, or giving your pictures to an organization for their marketing, get your license.

Register and get ID from FAA

All drones weighing more than one-half pound and less than 55 pounds, including the payload, must be registered. Go to FAAregisterdrone.com and follow the instructions. DO NOT get sucked into sites that charge a fee to file the paperwork. Registering your drone is very easy.

What is needed to register drones?

Registration costs $5 per drone and is valid for 3 years. To register, you'll need an email address; credit or debit card for the $5; physical address and mailing address (if different from physical address); and make, model and serial number of your drone. You must be 13 years of age or older. If the owner is less than 13 years of age, a person 13 years of age or older can register the drone. You must also be a U.S. citizen or legal permanent resident.

FAA restrictions on commercial drone operations

The FAA restricts commercial drone operations under Part 107 to drones weighing less than 55 pounds. Most commercial drones are around 25 pounds. The FAA has restrictions on commercial drone operations related to flights over people, at night, near airports, visual-line-of sight, below 400 feet AGL (Above Ground Level), faster than 100mph, avoiding manned aircraft, and flying from a moving vehicle. Waiver processes for each restriction are discussed in Lesson 66.

55

Always keep the drone in sight

The FAA requires that drone pilots always keep the drone within visual sight, which is approximately .8 miles. The FAA prohibits the use of binoculars to follow the drone. It is helpful to have a Visual Observer to watch the drone and look for any potential hazards while you operate your drone.

The categories that fly over people

To fly over people, you must have your Part 107 license and know what category your drone falls into. All four categories can fly over people if your drone follows safety guidelines and FAA regulations. These categories get specific on what your drone can do and offer options to get more permissions. If the manufacturer has not stated what category the drone falls into, the pilot must find their category.

How pilots can fly at night

If you have your Part 107 license you can fly at night. Your drone must have anti-collision lights that can be seen for three statute miles and have a flash rate that is sufficient to avoid a collision. Before you are able to do this the pilot must complete an updated knowledge test and recurrent training.

58

Check restrictions to fly near airports

Commercial drone operations need to have permission to operate in "controlled airspace," which is the area around airports, and restricted airspace. Technically, operations in Class B, C, and E airspace are allowed with the required ATC permission. Operations in Class G airspace are allowed without ATC permission.

… 59

FAA automated permission system at airports

In 2018, the FAA began to roll out Low Altitude Authorization and Notification Capability (LAANC), a collaboration between the FAA and industry. LAANC authorization to fly in controlled airspace is easy, digital, and available immediately. Authorization requests are submitted with a tap in the LAANC app and approved in seconds. Many drone mission software packages have integrated LAANC into the planning software.

60

Never fly drones near other aircraft

Drones must always give right-of-way to manned aircraft to avoid collisions. It is much easier for a drone pilot to see a manned aircraft than for a manned aircraft pilot to see a drone. Manned aircraft include big jets and smaller general aviation aircraft, helicopters, parachute jumpers, and even hot air balloons.

61

Never fly higher than 400 feet

The FAA has set the maximum altitude for commercial drones at 400 feet Above Ground Level (AGL) because general aviation aircraft and helicopters are at 500 feet and above. Many commercial drones are geofenced to not climb higher than 400 feet.

Never fly near emergency response efforts

As discussed in Lesson 7, Drone operations are very disruptive to emergency response efforts at times of natural and other disasters and are not allowed to be flown over disaster sites. These include wildfires and floods. Emergency response agencies can hire commercial drone operators under special exemptions to operate at times of emergency.

Check restrictions to fly national parks

The National Park Service has made it illegal to launch, land, or operate drones. But you should check with the park service. You might receive permission to fly in certain areas. The FAA has restricted drone flights above 10 national monuments. The FAA B4UFly app and the free app AirMap can provide the information you need to make sure your flight is legal.

64

Check local community restrictions to fly

Local communities are responding to privacy complaints from citizens and are passing rules restricting where drones can be operated. They also see an opportunity to make money with registration fees. For the most part, the FAA has the authority over drone operations, not local communities. However, local communities can restrict take-offs and landing in parks, but cannot restrict flying over a park, because the FAA controls airspace.

Requirements if drone causes human injuries

A commercial drone accident must be reported to the FAA within 10 days if there is (1) serious injury to any person or any loss of consciousness; or (2) damage to property (other than the unmanned aircraft) unless "the cost of repair (including labor and materials) does not exceed $500, or the fair market value of the property does not exceed $500 in the event of a total loss."

Getting FAA waivers from drone rules

The FAA has streamlined the application process for part 107 waivers and airspace authorizations. All applications must be submitted FAA.gov/DroneZone. Before applying, review the step-by-step application process. Waivers for operations in controlled airspace are regularly granted. Waivers for night-flying require a detailed safety plan for operating the drone without seeing the aircraft. Very few waivers have been granted for Beyond Visual Line of Sight (BVLOS).

Getting FAA exemptions from aviation rules

Any FAA regulation can be exempted if not required by law, such as operating a drone weighing more than 55 pounds. The request must include reasons the exemption would not adversely affect safety, or how the exemption would provide a level of safety at least equal to the existing rule. The FAA then publishes the exemption request in the Federal Register. The exemption process is much more difficult than waivers.

Best Practices for Safe Drone Operations

68

Hiring/providing drone services best practices

The Association for Unmanned Vehicle Systems (AUVSI), AUVSI.org, the largest organizations for drone operators, manufacturers, and service providers, has the Trusted Operator Program (TOP), which provides recommended best practices for a wide range of commercial drone operations. Individual pilots and training programs can be TOP certified. The following lessons discuss some key best practices.

Be considerate of invading personal privacy

Invasion of privacy by drones flying over and taking pictures of unsuspecting folks can be a major issue. A drone can pop over the horizon and fly near someone with its camera looking straight at them. There is nothing individuals can do to stop this, because it is a felony to shoot or disable aircraft, including drones. The only action is to call local law enforcement and have privacy laws enforced.

No insurance required but good idea

Insurance to operate a drone is not required by law, but flying without coverage exposes you to significant lawsuits. Drone insurance policies should include liability coverage for bodily injury, property damage, and personal injury (libel, slander, invasion of privacy). Drone insurance can be purchased as an annual policy or on a per-mission basis.

Check flight restrictions on drone apps

If you are thinking about a commercial drone project, checking for flight restrictions is easy. Load onto your mobile device the FAA B4UFly app and the free AirMap app to make sure you know flight restrictions in the area where you plan to fly.

Check weather status on drone apps

Commercial drone projects are dependent on good weather. Both those hiring drone operators and drone pilots need to appreciate that project deadlines need to be adjusted for weather. Weather dramatically impacts the performance of drones. In my opinion, the best weather app is UAV Forecast. It provides as very simple Go/No Go display, sunrise and sunset times, winds and temperature, possibility of precipitation, and the number of satellites available.

Special considerations when flying over water

When planning a project where the drone mission is over water, remember that reflections off the water can sometimes confuse the positioning sensing of the drone. Be sure to stay away from obstacles that could interfere with the drone's compass or signal, such as lighthouses, boat transceivers, antennas, and other electronic devices.

Importance of Advanced Drone Operations Training

Training is required by the FAA

After taking the Part 107 test, you must take the online training on the FAA website. This training takes 30 minutes to an hour and can be taken over multiple days. This is the only training required by the FAA, but they do offer more free training, such as flying at night. There is no practical test for a pilot's skills to fly a drone.

Training to pass the FAA test

There are many free training programs to help you pass the FAA Remote Pilot in Command knowledge test. Many are on YouTube. In addition, there are structured classroom-style programs and video-based programs for roughly $120.

Finding professional drone hands-on training programs

While a lot can be learned from YouTube and going out and flying, taking professional training for drone operations with a trained instructor is important to improve piloting proficiency and skills. Specialty training in areas such as search and rescue, bridge surveying, and cinematography are important to learn the best practices for these industries.

77

Safety is paramount in drone operations

It is important to promote a safety culture for companies hiring drone services and for drone operators. When planning operations, conduct a risk assessment to identify potential safety issues. Mission planning and checklists ensure a safe operation for all involved with flight operations.

Preparing Bids for Government Drone Contracts

Designing government requests for drone proposals

Federal, state, and local governments are recognizing the benefits of drone services to government programs. Government agencies looking for drone services should clearly define the mission requirements and qualifications expected from a drone company. Think through the various departments that might benefit from drone services. Once the Request for Proposals (RFP) is issued, expect proposals from drone companies all over the United States. My local government received 20 proposals!

A little schmoozing goes long way

Once an RFP comes out, most government agencies can no longer talk directly with bidders. It is a good idea to get to know potential government agencies before the RFP comes out. State and local divisions of public works typically issue the requests for drone services. Set up a meeting to get to know them.

Responding to government requests for proposals

For a small drone company, responding to a government RFP is time-consuming and very competitive. Each governmental entity has its own requirements for submitting a proposal. Be sure to follow the process outlined in the RFP. My proposal was disqualified because I did not use the exact form in the RFP.

Good strategy: go to pre-bid conferences

Pre-bid conferences give government agencies an opportunity to expand on the information provided in the RFP, gauge the interest level of responders and hear about any issues not addressed. For drone companies, pre-bid conferences provide opportunities to meet the decision-makers as well as competitors.

Question: to team or not team?

While at the pre-bid conference, bidders might decide to team up to provide a better spectrum of services. The RFP should clearly state how teaming will be handled by the government agency. Drone companies may decide to get together to share the burden of drafting the proposal and to expand the services being offered.

DBE, MBE, WBE, WOSB, VOSB, SDB

Industry and government encourage diversity with programs including Disadvantaged Business Enterprise, Minority Business Enterprise, Women Business Enterprise, Woman Owned Small Business, Veteran Owned Small Business, and Small Disadvantaged Businesses. The Small Business Administration certifies companies to meet certain criteria for each program. A first step in government contracting is to get your System for Award Management (SAM) number and create your profile.

Government can take long to pay

Government procurement processes take a long time and payment processing has many steps. As a small business, it is difficult to wait for payments from the government. There are banks that will advance money on government contracts, but they are expensive.

Finding Drone Companies and Selling Services

…

Hiring a drone company for projects

Your company has decided to hire a drone company for a project, now where can you find a good one? The following lessons discuss the many sources to find drone companies to fulfill specific missions and projects. This section also provides ideas for drone operators to find customers.

Create business plan; work business plan

Starting a drone company starts with creating a business plan and then working the plan. The business plan should include the market you plan to serve, skills and training, equipment, software, marketing strategy, and a financial plan. Start with a self-evaluation--do you have the time and capital to start a business? Do you have the personality to go knock on doors and get the business?

Finding customers for your new business

As you start your new drone business, you will need to conduct basic marketing. The drone industry needs to do more to help other industries understand the benefits of hiring a drone company. The book, *Six-Word Lessons for Successful Start-ups* provides helpful lessons for starting any company, including drone companies.

AUVSI good source for drone businesses

The Association for Unmanned Vehicle Systems (AUVSI), AUVSI.org, is the largest organization for drone operators, manufacturers, and service providers. The AUVSI Trusted Operator Program (TOP) lists TOP certified operators and is a good source for locating certified drone operators. The AUVSI TOP program provides drone operators lessons on the best practices for specific missions.

Facebook Groups can have useful information

There are several Facebook groups for commercial drones where ideas and techniques are shared. I personally follow Commercial SUAS Remote Pilots. There are other Facebook groups for companies looking for drone operations. Just search for drone services and local drone operators will be in the search results.

90

Drone pilot online marketplaces for services

There are a number of companies offering services to match companies looking for drone services to drone operators. These companies broker services, which helps companies find third-party drone pilots. The goal is to make it easier for companies to afford the expensive technology of drones and, in the process, help independent drone pilots find customers and make money. There are mixed reviews on these marketplaces for drone services.

Drone service business: get first customer

When starting a drone service business, it is easy to try to cover too many different types of customers. The best advice is to focus on getting your first customer by either having them come to you through social media, or you going to them. It is important to learn as much as you can about one type of client before blasting out to the world that your drone services are open for business.

Do your homework to win customers

Do your homework. For example, if you want to specialize in commercial or residential real estate, there is a lot of information on sites such as Realtor.com and Zillow.com. Find properties for sale and "do the job," fly the property, process the images, and create a sample customized marketing piece for the real estate listed. Deliver a professional product to the agent and open dialogue for future work.

Deliver services and products as professional

Professional presentation of marketing materials, proposals, and delivery of final products is important to customers. Acting professionally in the field, such as wearing a safety vest and following safety protocols, enhances your business and leads to referrals for more business.

Think outside the box for customers

Be creative when looking for a drone company to hire or for customers for your drone business. There are web services where customers can bid on all sorts of professional services. Even craigslist.com has a section for buying and selling drone services.

Free services do not generate customers

Most drone companies start out building a portfolio of services by taking on *pro bono* (free) assignments in the hopes that eventually the customer will pay for the drone services. I have found that *pro bono* customers never convert to paid customers. When you give the product for free, the customer does not value the service.

Creating a winning drone service pitch

Many potential customers need to be convinced of the benefits of using a drone. Demonstrating a drone's abilities by showing examples of images a drone can capture from the air is a great sales pitch. If you're not a natural salesperson, consider taking some sales training. Once you do start to land customers, you need to maintain sales and marketing efforts to get future customers.

97

Become part of the drone community

As a buyer or seller of drone services, it is important to stay up-to-date on developments in the drone industry. Technology, including aircraft, sensors, and software, is evolving at warp speed! And you need to keep an eye on federal, state, and local governments.

Pricing and invoicing for drone services

There are 6 elements to setting prices and invoicing for drone services: 1) the pilot's hourly rate; 2) drone and sensors; 3) prep work to get required permission to perform the mission, e.g. FAA waivers, 4); time it will take to complete the project; 5) insurance; and 6) post-production. Very experienced pilots with high-end equipment and pricey software charge more than a new pilot with a quadcopter and low-end software.

Try *Women And Drones* for resources

If your company is looking for a drone company, check out *Women And Drones* at WomenAndDrones.com, which encourages women to learn to fly drones and features really interesting stories about women drone pilots, like me! I was profiled in the *We Are All Awesome* project.

Drone publications and blogs can help

There are many drone publications, blogs, and eNewsletters. I really like Sally French's *The Drone Girl* blog at TheDroneGirl.com because she reviews different drones and other products, provides understandable explanations of government regulations, and has good postings on fun things to do with drones.

Contact Tulinda

For more information, or to share ideas, please contact:

Dr. Tulinda Larsen
CEO & Founder, Skylark Drone Research
TulindaLarsen.com
Tulinda.larsen@TulindaLarsen.com
Twitter: @TulindaLarsen

Tulinda is also the author of *Six-Word Lessons for the Drone Hobbyist*

References

Amerson, Alicia. *Six-Word Lessons for Drone Pilots and Outdoor Enthusiasts: 100 Lessons to Make Drone Flights Safe, Ethical and Green for Wildlife and Humans.* Pacelli Publishing. 2018.

Daniek, Maureen. *Six-Word Lessons for Successful Start-ups: 100 Lessons to Guide Your Company to Success.* Pacelli Publishing. 2013.

About the Six-Word Lessons Series

Legend has it that Ernest Hemingway was challenged to write a story using only six words. He responded with the story, "For sale: baby shoes, never worn." The story tickles the imagination. Why were the shoes never worn? The answers are left up to the reader's imagination.

This style of writing has a number of aliases: postcard fiction, flash fiction, and micro fiction. Lonnie Pacelli was introduced to this concept in 2009 by a friend, and started thinking about how this extreme brevity could apply to today's communication culture of text messages, tweets and Facebook posts. He wrote the first book, *Six-Word Lessons for Project Managers*, then he and his wife Patty started helping other authors write and publish their own books in the series.

The books all have six-word chapters with six-word lesson titles, each followed by a one-page description. They can be written by entrepreneurs who want to promote their businesses, or anyone with a message to share.

See the entire ***Six-Word Lessons Series*** at **6wordlessons.com**

www.ingramcontent.com/pod-product-compliance
Lightning Source LLC
Chambersburg PA
CBHW070643050426
42451CB00008B/281